The INSIDE GUIDE

STEM CAREERS

Jobs in Robotics

By Beatrice Harris

Cavendish Square

Published in 2024 by Cavendish Square Publishing, LLC
2544 Clinton Street Buffalo, NY 14224

First Edition

Library of Congress Cataloging-in-Publication Data

Names: Harris, Beatrice, author.
Title: Jobs in robotics / Beatrice Harris.
Description: Buffalo, New York : Cavendish Square Publishing, [2024] |
Series: The inside guide: STEM careers | Includes index.
Identifiers: LCCN 2023032942 (print) | LCCN 2023032943 (ebook) | ISBN 9781502670434 (library binding) | ISBN 9781502670427 (paperback) | ISBN 9781502670441 (ebook)
Subjects: LCSH: Robotics–Vocational guidance–Juvenile literature. |
Robots–Juvenile literature.
Classification: LCC TJ211.25 .H38 2024 (print) | LCC TJ211.25 (ebook) |
DDC 629.8/92023–dc23/eng/20230809
LC record available at https://lccn.loc.gov/2023032942
LC ebook record available at https://lccn.loc.gov/2023032943

Editor: Therese Shea
Copyeditor: Shannon Harts
Designer: Deanna Lepovich

The photographs in this book are used by permission and through the courtesy of: Cover, pp. 4, 12, 13, 20, 21 Gorodenkoff/Shutterstock.com; p. 6 Stefano Mazzola/Shutterstock.com; p. 7 (main) Shotmedia/Shutterstock.com; p. 7 (inset) Olga Miltsova/Shutterstock.com; p. 9 Milan Ilic Photographer/Shutterstock.com; p. 10 goodluz/Shutterstock.com; p. 14 Factory_Easy/Shutterstock.com; p. 15 worradirek/Shutterstock.com; p. 16 kung_tom/Shutterstock.com; p. 18 NASA; p. 19 Arsenii Palivoda/Shutterstock.com; p. 22 Triff/Shutterstock.com; p. 24 KlingSup/Shutterstock.com; p. 25 MikeDotta/Shutterstock.com; p. 26 oakstudio/Shutterstock.com; p. 27 Ground Picture/Shutterstock.com; p. 29 (left) lokanan VFX Studios/Shutterstock.com; p. 29 (right) VesnaArt/Shutterstock.com.

CPSIA compliance information: Batch #CWCSQ24: For further information contact Cavendish Square Publishing LLC at 1-877-980-4450.

Printed in the United States of America

CONTENTS

Robots aren't always humanlike. These factory machine arms are robots, for example.

Chapter One

READY FOR A ROBOTICS CAREER?

Few inventions inspire our imaginations like robots can. We can't help but picture how they could aid us and how they could change our lives. Robots are already changing lives and are operating in a number of industries, including manufacturing, space, and medicine, as well as more surprising places. For example, are you ready for a robot teacher?

Most of us don't have day-to-day contact with **complex** robots, but we probably will soon. Robots are replacing more and more of the tasks that people do. In the future, they'll likely deliver packages to your door, greet you in stores, and even drive you places.

Fast Fact

A company called Miso Robotics created a robot that has been programmed to recognize, or sense, certain kinds of foods and how to cook them.

Defining a Robot

What's the difference between a robot and a machine? While there are many different definitions, in general, a robot is a kind of autonomous machine, which means it can operate by itself. Levels of autonomy can vary, but all robots need programming, or instructions in a computer language. Machines, on

Have you been served by a robot in a restaurant yet? This industry is testing out robots in many positions, including chefs.

> **Fast Fact**
>
> The robot's controller, a kind of computer, uses **software** and data collected from **sensors** to make decisions, such as whether to have a robotic part perform an action.

the other hand, require a lot more hands-on human operation.

Robots are made to do actions that a person would usually do. Many of these tasks are **repetitive** or unsafe. While robots require people to program them, they often can be left to their tasks. More autonomous robots can even change their

This robot vacuum uses sensors to make a decision about whether to move forward or backward.

actions according to their **environment**. A simple way of defining a modern robot is that it's a machine that can sense, calculate, and act.

Robotics is the field of science that has to do with the design, construction, programming, and use of

While a machine such as a remote-controlled car can act, it can't sense and calculate like a robot can.

WORRIES ABOUT ROBOTS

Robotics is sometimes a controversial field. That means people have strong feelings about it and often disagree. Some of that controversy comes from the worry that robots will take away people's jobs. For example, people once did some of the jobs in factories that robots now handle. It's true that robots are taking some kinds of jobs from people, but experts report they're creating more and different kinds of jobs. According to the World Economic Forum Future of Jobs Report 2020, while 85 million jobs may be shifted to machines (including robots) by 2025, 97 million new jobs may be created.

robots. It is many areas of knowledge coming together to create an advanced machine. Robotics careers include people who plan the robot, or decide what it will look like and what its purpose is. People are also needed to build the robot, including all its "body" parts and the electronics that run through it. Robotics also involves programming—creating the instructions the robot needs to complete tasks. Finally, some people who work in robotics operate, oversee, and maintain the robot. The field of robotics has many job opportunities!

Fast Fact

Robots can lift heavy objects, touch hot substances, and do work in places too dangerous for people, such as underground mines and deep water.

Cutting-Edge Careers

Robotics requires a great deal of STEM knowledge. That's probably a term you've heard at school. STEM stands for science, **technology**, engineering, and math. All these fields are well represented in the careers you'll learn about in this book. For example, math is important in every part of a robot's construction, and math is used in programming too.

As you read, think about what job might suit your interests and talents. Since robots are spreading into so many areas, reflect on what you'd want to develop a robot for. This cutting-edge career is full of exciting possibilities!

Some toys teach young people about the basics of robotics. They're a great way of getting a start on a robotics career.

A robotics engineer and technician team up to work on an autonomous drone, which is an aircraft without a pilot.

THE BUILDERS

Chapter Two

Robotics engineers are leaders in the creation of robots. They must have an understanding of several kinds of engineering fields. In fact, some robotics engineers focus on one kind of engineering and make their way into the robotics field successfully. However, a growing number of colleges offer a robotics engineering degree that combines engineering subjects.

Robotics Engineers

One of the first steps engineers take at the beginning of a new robotics project is to meet with the people who will use the robot. That could be a business owner, military officials, or someone else. The engineer discusses with them what the robot's tasks will be and other details, such as size, power source, and other concerns. Then, the engineer begins to design, often using computer-aided design (CAD) software. This software not only helps them sketch what the robot might look like, but also shows how it would move. They'll usually

Fast Fact

Training courses called boot camps are a way to learn certain robotics skills, such as coding.

Here, CAD software helps engineers design a robotic dog. The software provides realistic, or lifelike, images.

Fast Fact

Hydraulics is the use of liquid pressure to move mechanical parts. Pneumatics uses air or gas pressure to move mechanical parts.

work with a team to come up with a design that everyone agrees on.

Next, they'll make a prototype, or a model. They often use computer-aided manufacturing (CAM) software to guide the machines that make many of the parts. The prototype helps the robotics engineer test the whole system—the hardware and software. Depending on their level of knowledge in each area, they

BASIC ROBOT PARTS

All robots have some basic parts. First, the controller functions as the robot's brain. It's basically a computer that uses software that can deliver commands to the robot. The robot also has mechanical, or moving, parts that carry out the controller's instructions. These parts include wheels and end effectors, or tools that interact with the environment. Electrical motors may drive the moving robotic parts, or another power source may, such as pneumatics and hydraulics. Sensors connected to the controller provide data so that the robot can make adjustments to its instructions. Sensors may include microphones to sense noise, thermometers to sense temperature, and cameras to sense obstacles and distance.

This robot's end effector is called a gripper.

may work with electrical engineers, mechanical engineers, computer engineers, and others to fix problems that arise. After the production of a successful prototype, the robotics engineer may oversee further production as well as perform maintenance on older models.

Robotics Technicians

Once the robot is built, it's sent where it's needed. Like any machine, it needs to be repaired and adjusted from time to time. The professional to do that is likely a robotics technician.

Robotics technicians in manufacturing plants often see to the mechanics of robots. They need an understanding of hydraulics, pneumatics, and electronics, as these are responsible for the movements

Technicians may swap a robot's end effector, or end-of-arm tool, for a different one. End effectors include grippers and welding and painting tools.

of many robots. In addition, they may have to understand the software side of the robot, particularly if a problem arises with its programming. They may have to call in another, more specialized professional to fix it, however.

Robotics technicians may also work with robotics engineers in the construction of a robot, helping test it at various stages. So, they need an understanding of the engineering involved in robotics.

Fast Fact

The devices that make a robot's parts move are called actuators.

Most robotics technicians have graduated from high school and obtained at least a two-year degree, called an associate's degree, in a robotics field. They may also go through a special training program before joining the workforce.

One of a robotics technician's most important skills is the ability to recognize where the problem is when a robot isn't functioning properly.

Robotics software applications allow robots to do a variety of tasks on farms.

Chapter Three

SOFTWARE EXPERTS

A robot wouldn't be a robot without the software driving its computerized "brain." The creators of this software are among the most highly paid professionals in the robotics field. This is an exciting branch of robotics that's making remarkable advancements.

Software Engineers and Developers

Some companies consider robotics software developers and robotics software engineers to be the same. However, developers often work on part of an application, while engineers work on a whole software program or system. Every robot created so far, no matter how autonomous, has needed programming to function. Robotics software engineers develop or adapt software to control robots and provide them with autonomy. Most don't start from scratch but use software that's already available.

Fast Fact

ROS software and tools are open source, which means they're available for anyone to use and change.

For example, the Robot Operating System (ROS) is a collection of software and tools on which to build a robot application. Many robotics companies use ROS or the improved ROS 2. Both provide a framework that engineers can start from.

The NASA robotic system called Astrobee was developed using ROS. The cube-shaped robot Bumble, which is part of Astrobee, is shown here on the International Space Station.

Fast Fact

Software engineers use different kinds of advanced math, including calculus, statistics, and algebra.

When software engineers are done creating the software that connects the robot's actuators, sensors, and controller, they do a lot of testing and debugging, which means finding bugs, or mistakes in code, and correcting them.

Robotics software engineers usually must have a college degree in computer science, engineering, or a similar field. They should be skilled in using computer languages such as Python, C++, and Java.

Artificial Intelligence Specialist

Artificial intelligence, or AI, is a branch of computer science in which computers are programmed with data that allows them to make decisions and problem-solve like people can. Because robots have computers—in the form of their controllers—they, too, can have artificial intelligence software.

AI specialists program robots to simulate, or imitate, human intelligence. The basis of such an AI system is a lot of data and complex algorithms. An algorithm is a series of steps a computer can follow to solve a problem. Some algorithms can recognize patterns in data to **predict** values. When new data is provided to these algorithms, they learn and improve their performance. This is called machine learning.

If you've ever talked to Amazon's Alexa or communicated with an online **chatbot**, these are examples of AI systems. They predict what

Artificial intelligence specialists may have other job titles depending on their focus. These include machine learning engineer and computer vision engineer.

LEVELS OF AI

Not every AI system is as smart as another. There are three levels, or stages. The first, called artificial narrow intelligence (ANI), is considered the most basic, as it allows machines to do only one function. Examples are a robot that can play chess, but can't "learn" to do anything else. Stage two is artificial general intelligence (AGI), in which the computer can solve problems for a variety of tasks and reason as well as humans. Some think robotics professionals are just starting on stage two. The third stage is artificial super intelligence (ASI). This would be a computer system or set of systems that outperforms human intelligence across all fields of knowledge.

Research in AI robotic legs is giving hope to people who have difficulties walking.

AI specialists can use frameworks for their programming, just as other areas of robotics use ROS.

the user wants. The more data the user (and programmer) provides, the more the AI system learns.

Generally, AI specialists have a bachelor's degree in computer science or a related field, and many have higher degrees as well. They're skilled in programming languages including Python, Lisp, and Java.

Fast Fact

Artificial intelligence specialists find work in the research departments of universities as well as AI software and robotics companies.

Hundreds of scientists, researchers, engineers, and other professionals worked on NASA's robotic exploration vehicle that landed on Mars in February 2021. This robot is called a rover.

OPERATING, DESIGNING, AND SELLING ROBOTS

Chapter Four

If you're not interested in building and programming a robot, it's not the end of the road for your robotics career. There are still more jobs that might better match your interests and goals. Check out some of these interesting careers!

Robot Operator

After a robot is delivered to where it will do its work, it needs someone to oversee it. That might be the responsibility of a worker called a robot operator. This person's job duties can vary, but may include preparing the robot for operation, maintaining it, and making sure the robot is operating successfully. They also might use preprogrammed software to make adjustments as needed.

A robot operator might not need a college degree. Robotics manufacturers may provide them with the training they'll need on the job.

> **Fast Fact**
>
> Those with an interest in robotics and cars can combine their interests while working to develop fully autonomous vehicles.

UI or UX Designer

People want to work with robotics systems that are easy to operate.

A UX or UI designer tries to understand how an operator might interact with a robot.

Fast Fact

In robotics, UX and UI designers work together on a team with engineers and developers.

Also, in many situations, especially if the robot is interacting with the public, the robot should look friendly—not intimidating, or scary. The professionals that work on these issues are called user **interface** (UI) designers or user experience (UX) designers. In general, UI designers focus on the appearance of computer systems and software interfaces, while UX designers focus more on a user's overall experience and how the user interfaces work. UI and UX design may be combined into one role.

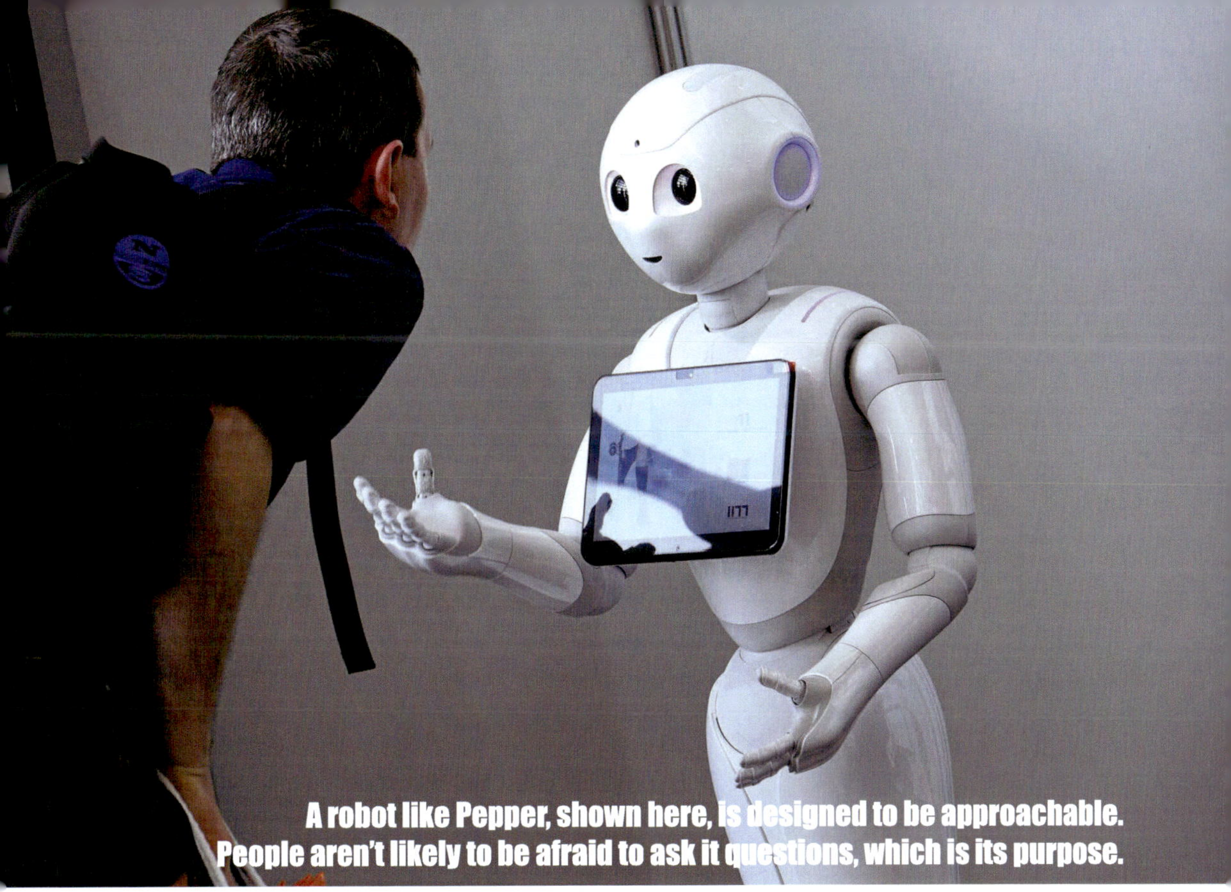

A robot like Pepper, shown here, is designed to be approachable. People aren't likely to be afraid to ask it questions, which is its purpose.

Employers may be looking for people with a bachelor's or more advanced degree in UX or UI design or a related field. Experience is especially helpful in getting a job in these areas too.

Sales Representative

Robotics companies need people to sell their robots, robotics systems, software, and related products. A robotics sales representative is a perfect job for people who know a lot about robots and how they can help others. Sales representatives need to have some knowledge of robotics technology, but also must have great "people skills." They need to be able to come to a business agreement with customers and solve problems when things go wrong.

NANOROBOTICS

A nanometer (nm) is a measurement equal to one billionth of a meter. How small is that? A human hair is about 60,000 nanometers thick! A branch of robotics focuses on building robots on the nanoscale, or up to 100 nanometers. Researchers are hopeful that robots this small could have a major impact in the medical field. Nanorobots could, for example, deliver medicine to a target within a human body. They also could be used in manufacturing to create materials with fewer imperfections. Some think they could repair and maintain spacecraft too. As of 2023, nanorobotics is still in the research stage.

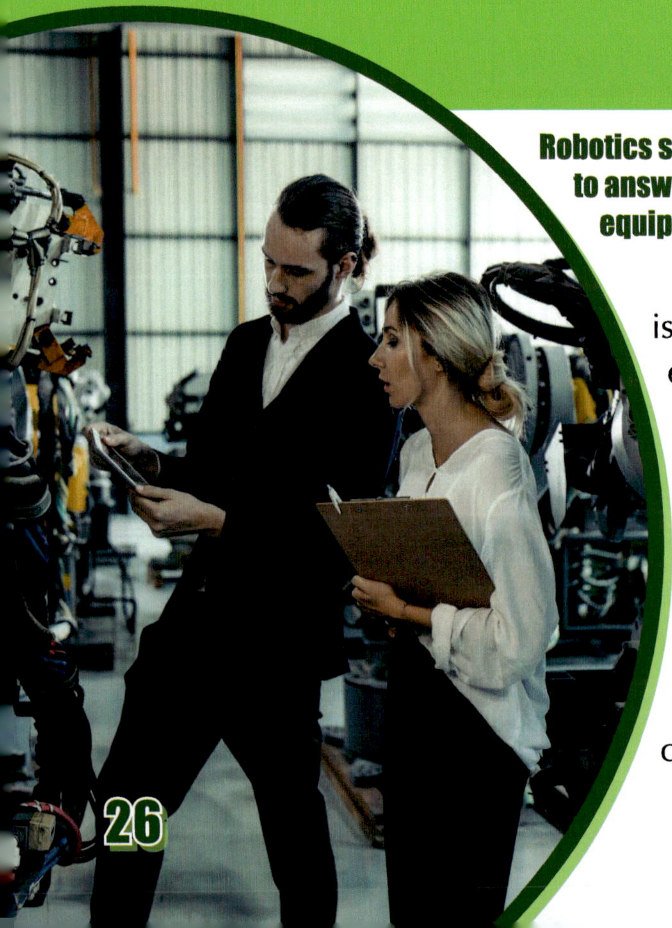

Robotics sales representatives should be able to answer questions about the robotics equipment they're selling.

For this career, a college degree isn't necessarily required, but sales experience is a must.

Ready for the Future?

People who are attracted to robotics careers have similar interests, whether they're a robotics engineer or a sales representative. They're confident about this technology's place in our world and excited to be a part of the advances that will come.

To prepare to join these robot enthusiasts, you can study hard in school, especially in your STEM classes. Join a school, library, or community robotics club. You'll discover a lot about the basics of robotics engineering in a group like this. You'll also meet new friends and have fun!

Fast Fact

More schools are offering robotics classes, so more robotics teachers will be needed at all levels of education.

Many robotics clubs have competitions with other clubs as a way to test robots' abilities.

MORE PLACES TO FIND ROBOTS

dairy farms
milk cows and monitor their health

hotels and hospitals
transport needed items to people in their rooms

medical school
act as human subjects for medical students to learn procedures on

military
transport supplies, perform search-and-rescue missions, and clear **minefields**

***FIRST* Robotics competitions**
compete in local, regional, and national robotics events

sports
collect tennis and golf balls and act as officials in some baseball games

THINK ABOUT IT!

1. Are you fearful of robots with artificial intelligence? Why or why not?
2. Other than the places mentioned in this book, where else could robots be useful?
3. Which jobs should not be replaced by robots, and why?
4. Which career in this book was the most interesting to you, and why?

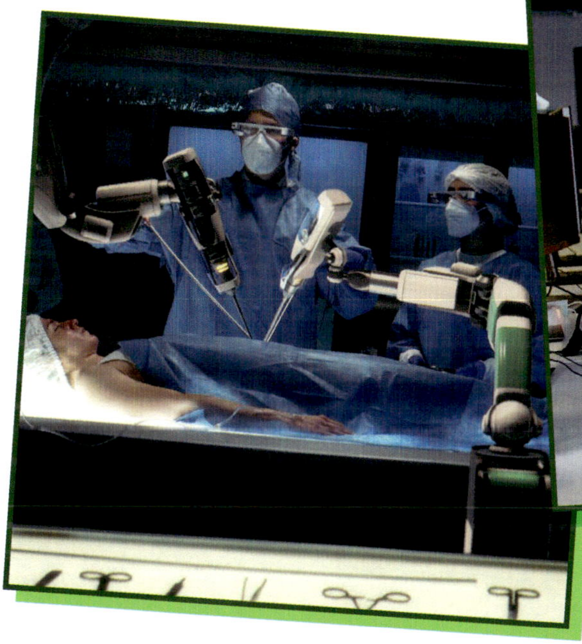

GLOSSARY

application: A program that is made to perform certain functions, or tasks.

chatbot: A computer program designed to communicate with people.

complex: Having to do with something with many parts that work together.

environment: The conditions that surround someone or something.

interface: A program that controls how information is shown to a computer user and how they can communicate with the computer.

minefield: An area where explosives are hidden in the ground.

predict: To state an outcome in advance based on known data.

repetitive: Repeating many times.

sensor: A tool that can detect changes in its surroundings.

software: A computer code, or program, that instructs a computer about what to do.

technician: Someone whose job has to do with the use of machines.

technology: A method that uses science to solve problems and the tools used to solve those problems.

weld: To join metal parts together by heating.

FIND OUT MORE

Books

O'Brien, Cynthia. *Where in the World Can I . . . Design a Robot?* Chicago, IL: World Book, Inc., 2023.

Virr, Paul. *The Brainiac's Book of Robots and AI*. New York, NY: Thames & Hudson Inc., 2023.

Zajac, Linda. *Robo-motion: Robots That Move Like Animals*. Minneapolis, MN: Millbrook Press, 2022.

Websites

Robotics: Facts
sciencetrek.org/sciencetrek/topics/robots/facts.cfm
Read a short history of robots, and find out other interesting robotics information.

ROS—Robot Operating System
www.ros.org/
Learn more about ROS through an informative video.

What Is a Robot?
robotsguide.com/learn/what-is-a-robot
This is a great summary of the work that goes into building a robot.

INDEX